AN EASY APPROACH TO ACCEPTANCE SAMPLING —
HOW TO USE MIL-STD-105E

AN EASY APPROACH TO ACCEPTANCE SAMPLING — HOW TO USE MIL-STD-105E

Richard T. Weber

ASQC Quality Press
Milwaukee, Wisconsin

AN EASY APPROACH TO ACCEPTANCE SAMPLING —
HOW TO USE MIL-STD-105E

Richard T. Weber

Library of Congress Cataloging-in-Publication Data

Weber, Richard T.
     An easy approach to acceptance sampling: how to use MIL-STD-105E / Richard T. Weber.
     p.   cm.
     Includes bibliographical references and index.
     ISBN 0-87389-118-X
     1. Acceptance sampling. I. Title.
TS156.4.W43 1991
658.5'62—dc20
91-14599 CIP

10987654

ISBN 0-87389-118-X

Acquisitions Editor: Jeanine L. Lau
Production Editor: Mary Beth Nilles
Set in Clearface by DanTon Typographers. Cover design by Artistic License.
Printed and bound by BookCrafters.

For a free copy of the ASQC Quality Press Publications Catalog, including ASQC membership information, call 800-952-6587.

Printed in the United States of America

American Society for Quality

Quality Press
611 East Wisconsin Avenue
P.O. Box 3005
Milwaukee, Wisconsin 53201-3005
414-272-8575
Fax 414-272-1734
800-248-1946
Web site http://www.asq.org

CONTENTS

PREFACE

Although MIL-STD-105E is widely used in both commercial and military businesses, those who use it generally do not understand it. This leads to waste due to poor systems and poor decisions. The training method and procedures contained in this book have been used by more than 500 inspectors, engineers, and managers with gratifying results. After participating in the training, managers develop better systems. Inspectors use their information to improve quality decisions. Engineers are better able to assess process performance and prepare improved inspection instructions.

If your business continues to use inspection by attributes, a training program is recommended. This book provides a training method that is easy to understand. This training method was developed to show business managers and inspectors with various levels of statistical knowledge the decisions being made while using MIL-STD-105E.

Many methods of predicting process variation exist. One method that has been widely used is inspection by attributes. Some have proposed that this inspection technique results in a lack of quality control. Inspection by attributes does not result in a lack of control. It is the lack of application and the attitudes of those who use it that cause poor quality.

The basic concepts of inspection by attributes have been corrupted over the years. The acceptable quality level (AQL) does not mean that a certain level of defects is acceptable. Use of this definition indicates a poor attitude toward quality. AQL should be defined as a process average. A process is measured to determine its specific level of quality, and if that level is satisfactory, subsequent measurements are then taken to ensure that the quality level has not changed statistically. Thus, the inspection by attributes technique is pro-active and can be used to make decisions.

Just like control charts and other statistical tools, a rejection based on this inspection method indicates the process shifted or the rejection was false. The rejection of a lot should drive an effort to determine whether the results were valid. Lot acceptance indicates the process average has a probability of fitting within a set range. Process improvements are made by investigating rejected lots, finding what caused the rejection, and implementing change to improve the process. It is true that large sample sizes are needed to predict the quality with a small interval. However, many other statistical techniques need a large sample size to predict the population accurately.

The advantage of MIL-STD-105E comes from the tables and charts contained within the document. These tables and charts are tools to assist with measuring and controlling process performance. This book will provide an understanding of the inspection by attributes approach and MIL-STD-105E. It will also provide a visual method for demonstrating the technique to those who will use it. Perhaps with a better understanding of the technique, the inspection method will continue to be widely used to control processes and quality.

1
INTRODUCTION

The use of inspection sampling by attributes, MIL-STD-105E, in manufacturing and other quality systems continues even though it does have limitations. Some companies use it because their government and government-related contracts require attribute sampling to be structured from MIL-STD-105E. Others use it because their customers use it. Some companies use it because they think they know how to apply it. Others have evaluated the standard's capabilities, performed in-depth analyses, issued instructions, completed training, and applied it for performance improvement. Properly developed, used, and understood, MIL-STD-105E has its niche in manufacturing and other environments where sampling on a go/no-go basis is needed.

Some companies have modified the standard by developing their own inspection plans, by performing 100 percent inspection, or by using inspection plans with an acceptance number of zero. The latter sampling plan accepts the lot only if there are no defects in the sample; the lot is rejected if one defect is found, regardless of the sample size.

Companies that apply the standard without developing a sampling approach and providing training cause more harm to themselves than might be expected. Companies must understand sampling theory and know of special considerations when developing a sampling approach. They must also understand what is happening during attribute sampling inspection.

Few companies realize the associated risk of using MIL-STD-105E: Defective parts could be passed into the system. One common mistake involves interpreting the amount of protection a certain AQL provides. For example, ask an inspector or quality engineer the question, "Does a 4 percent AQL more often reject or more often accept a lot having 4 percent defective material?" Unless there has been a recent training program, at least half will not know, another 25 percent will give the wrong answer, and, if luck prevails, the rest might give the right answer. It is important to understand the results so that a good decision can be made. This understanding will also assist in evaluating why defective material was moved to the stockroom or from one operation to the next. Outside of quality control, few individuals would be able to correctly answer this question. As far as the management, materials, and manufacturing functions are concerned, the material was inspected, therefore it is good.

Besides understanding what the sampling risk is, consideration should be given to the amount of money being spent on inspection by sampling. Valuable inspection time might be used without meaningful results. A large amount of the quality department budget might be spent on inspection labor. If the labor is being misapplied because inappropriate sampling plans are used, then money is being spent for nothing. If the sampling plans are improperly developed, the inspectors might not be following them because the plans are impractical.

This book provides a background on sampling by attributes. It also provides a visual method for demonstrating sampling inspection by attributes; procedures for developing inspection plans and describing sampling inspection by attributes; definitions of terms; and sample inspection plans developed from MIL-STD-105E.

2
OVERVIEW—ACCEPTANCE SAMPLING BY ATTRIBUTES

Acceptance sampling by attributes in accordance with MIL-STD-105E is based on several primary concepts. These concepts are listed as follows:

1. The AQL is designed to most often accept lots of material having defect percentages equal to the AQL value.

2. The process average was measured and is stable and acceptable.

3. The part is either good or bad. In the case of evaluating a single complex product, the characteristic is either acceptable or unacceptable.

4. The defects are randomly distributed, and the sample is taken randomly.

5. The probability of finding a defect stays the same as samples are taken.

Good or Bad

Inspection by attributes means the product is classified as either defective or nondefective. For example, Figure 2.1 shows three similar parts. The inspection of these parts involves measuring the length, width, height, and weight of each part compared to a requirement. *The part is defective if one characteristic or all the characteristics are defective.* If the X dimension is out of tolerance in part a, the X and Y dimensions are out of tolerance in part b, and all the characteristics are acceptable in part c, there are two defective parts. The fact that part b had two defects does not vary the number of defective parts.

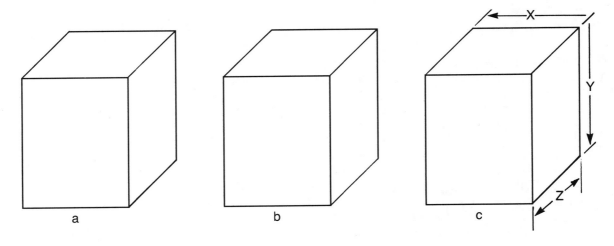

FIGURE 2.1

In the case where the number of defects within a product is counted, the characteristic is classified as a defect or a nondefect. The characteristic is either good or bad.

Random Sample

Paragraph 4.5.1 of MIL-STD-105E requires that samples be randomly selected. A random sample is obtained by using a random number table. Each part is assigned a random number and is entered into the random number table in a nonsystematic way. Then, the column is followed horizontally or vertically until a sufficient quantity is obtained to match the sample size. The part assigned the particular number from the random number table should be chosen.

If a lot of material has all the defects in one location, several conditions could have occurred. For instance, all the defective parts could have been made at the beginning of the run, all the defective parts could have been put in the bottom or top of the container, or the defective parts could have been sorted from the good but were accidentally put back in the lot.

Formation of the Lot

When forming or inspecting a lot or batch, paragraph 4.3 of MIL-STD-105E requires that the lot or batch consist of units of products of a single type, grade, class, size, or composition and that the lot or batch be manufactured under essentially the same conditions and at essentially the same time. If sampling inspection is used, appropriate capability studies and process development are required to ensure these parameters are maintained.

When presenting a lot or batch to the government for inspection, it must be properly identified and kept in adequate and suitable storage.

In actual applications, many of the conditions required for forming a lot might not be known or controllable. For example, receiving inspectors might not be able to control whether the parts they inspect are from a continuous process. Unless there are well-established supplier process requirements, the receiving inspectors proceed without the needed process knowledge.

The main purpose of receiving inspection is to protect the next process or product from receiving defects. This objective is not the same as assessing whether a process has shifted. The question for the receiving inspector is, "What AQL will protect the next process or part from receiving defects?" If the answer to this question is different than the capabilities of the process being evaluated, rejections might occur.

AQL—Process Average—Operational Characteristic Curve

MIL-STD-105E's definition provides that the process average be known, determined by a process capability study or similar study. The standard's tables are designed to match a theoretical curve shown in Figure 2.2. The concept of the AQL sampling procedure is to accept material that has a process average less defective than the AQL level. This infers that the process percent defective is known and a series of lots from the same process will be inspected.

The theoretical curve is interpreted as follows: A lot of parts that has a defect percentage equal to or less than the AQL percent defective will always be accepted. A lot that has a defect percentage greater than the AQL percent will always be rejected. In actual applications, the lot will probably be accepted if it contains a percent defective equal to the AQL. This theoretical probability curve is designed so that 90 percent to 95 percent probability of acceptance occurs at the AQL level. An example of an actual curve, referred to as an operational characteristic (OC) curve, is superimposed on the theoretical curve shown in Figure 2.3. The shape of the OC curve is dependent on the sample size and accept number. As the sample size increases for a particular AQL, the shape of the OC curve more closely matches the theoretical curve of Figure 2.2. As can be seen from the curve, there is a very good chance that a high percent of defective lots will be accepted. A rejection indicates the process produced an abnormally large number of rejects. A series of lot rejections indicate the process has a higher percent defective than the set AQL level—the process changed. An investigation and corrective action are needed to return the process to original quality levels.

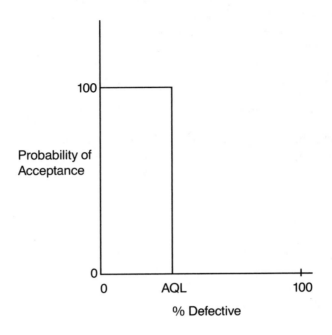

FIGURE   2.2

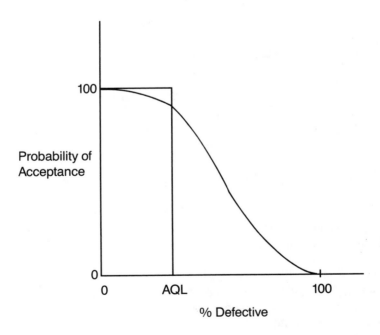

FIGURE   2.3

Two limiting quality tables in the standard give the percent defective for a 10 percent and 5 percent probability of acceptance. These values are commonly referred to as consumer's risk values. The tables can be used to determine the worst-case quality level likely to be accepted with the sampling plan being used. For isolated lots, the 50 percent probability-of-acceptance point and its related percent defective can be used as the most likely percent defective in the lot. The 50 percent probability of acceptance represents the most likely percent defective accepted on a lot-by-lot basis. The selection of an AQL might indicate the process or assembly using the parts from the sampled lot can tolerate parts with percent defective as high as the AQL or higher. The user's risk of accepting defective material can be determined from the tables in MIL-STD-105E.

In the case where the accept number is zero and the reject number is one, the shape of the OC curve is steep for low percent defectives, as shown in Figure 2.4. This can be further observed by comparing the probabilities shown in the tables of MIL-STD-105E. Some companies use these plans to control defects because the accept-on-zero curve has such a steep slope for low percent defectives.

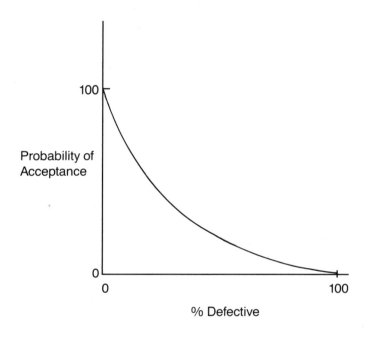

FIGURE   2.4

The Probability of Finding a Defect

For sample sizes of 80 or less, MIL-STD-105E uses the binomial distribution to calculate the probability of finding a defect. For sample sizes above 80, the standard uses the Poisson distribution. Both of these math models assume the probability of finding a defect stays the same as a sample is removed from the population.

An example of where the probability varies can be demonstrated with a deck of cards. In a regular deck, there are 52 cards bearing one of four possible suits. Each suit has 13 cards of varying value. The deck, therefore, has four cards of each value. The chance of randomly selecting an ace out of the deck is four out of 52 (one chance in 13). If one ace is removed from the deck, the chance of randomly selecting an ace out of this new deck is three out of 51 (one chance in 17). The probability of finding the ace changed. If the ace was the defect, the chance of finding the defect changed.

Similarly, if a card other than the ace was removed permanently from the deck, the probability of finding an ace would be four out of 51. This difference is not as dramatic compared to four out of 52, yet the condition is not a true binomial. Where the probabilities of finding a defect vary as defects are removed, the probability distribution is hypergeometric.

During inspection sampling, thought should be given to the sample's effect on the probability of finding a defect. If dramatically affected, a method that returns the sample to the lot, followed by a randomizing mixing, should be chosen.

Count Errors

During some inspection activities, physical counts are made to determine the lot size and subsequent sample size. In other cases supplied information, such as receiving paperwork or bills of material, are used to set the lot size. Count errors can easily occur and can result in:

- An improper sample size.

- A record error. This might affect inventory counts in cases where inspection counts are used for inventory purposes or to verify received quantities.

- A customer audit finding.

- A lack of traceability of critical material and subsequent hold of product shipment where traceability and configuration control are required.

Inspection Levels

In MIL-STD-105E, there are three inspection levels (I, II, and III) and four special levels (S-1, S-2, S-3, and S-4). Inspection level II is normally used. In accordance with paragraph 4, inspection level I can be used when less discrimination is needed; level III can be used when greater discrimination is needed. The S-1, S-2, S-3, and S-4 levels are used when relatively small sample sizes are needed and large sampling risk can be tolerated. These special levels are especially helpful when the cost of sampling is high, when destructive inspection is occurring, and when process and product instability can be determined from a small sample.

From an application standpoint, operational characteristic curves can help determine which sampling plan to use. They make it easier to picture the risk associated with the sampling technique. In the standard, the OC curves are displayed for the AQLs referenced.

Normal—Tightened—Reduced

The standard and its inspection tables are designed to detect a process that has varied from its measured performance. As a result, the normal inspection table is generally used first. Some people use the reduced inspection table first to achieve a smaller sample size. Reduced inspection plans have associated high risks because the standard generally uses a looser AQL when specifying reduced inspection. If smaller sample sizes are desired, they can be achieved by selecting a special inspection level instead of the reduced tables. If a government contract exists, paragraph 4.6 requires that inspection level II be used at the start of the inspection process before reduced or tightened inspection is invoked.

Four criteria must be met when switching from normal to reduced inspection:

1. Ten consecutive lots must be accepted by normal inspection during the original inspection. This means the lots cannot be previously inspected or rejected, and no special sorting can be done to any of the lots.

2. The total number of defects in the sample from those 10 lots must be equal to or less than the applicable number given in Table VIII of MIL-STD-105E. The left column of Table VIII identifies the number of sample units inspected from the last 10 lots or batches. Using a 4 percent AQL as an example, if the number of samples in each lot inspected was 50, the total number of samples in the last 10 inspections is 500. (It is possible the sample size would vary from lot to lot in which case the sum of the samples would be totaled.) The row that contains the quantity of 500 intersects the column headed by "4" at number 14. This means that only 14 or fewer defects can be found in 10 consecutively accepted lots if the switch to reduced inspection is to be made. If 15 or more defects are found, reduced inspection cannot be used.

3. Production must be at a steady rate.

4. Reduced inspection must be considered desirable.

Switching from reduced to normal inspection occurs when a lot is rejected. To return to reduced inspection, the criteria for switching to reduced inspection must be met again.

Switching from normal to tightened inspection occurs when two out of five consecutive lots are rejected. Switching from tightened to normal inspection occurs when five consecutive lots are accepted.

Charts showing the flow from normal to reduced inspection and normal to tightened inspection are provided in Figure 2.5 and Figure 2.6, respectively.

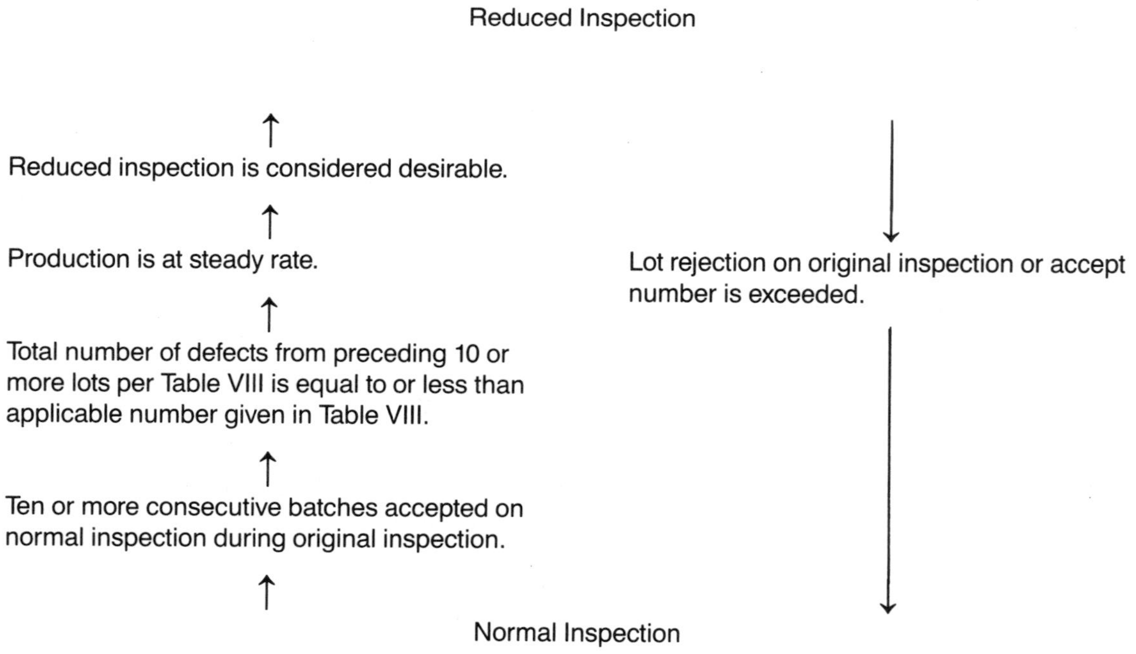

Reduced Inspection

↑
Reduced inspection is considered desirable.

↑
Production is at steady rate.

↑
Total number of defects from preceding 10 or more lots per Table VIII is equal to or less than applicable number given in Table VIII.

↑
Ten or more consecutive batches accepted on normal inspection during original inspection.

↑
Normal Inspection

↓
Lot rejection on original inspection or accept number is exceeded.

↓

FIGURE   2.5

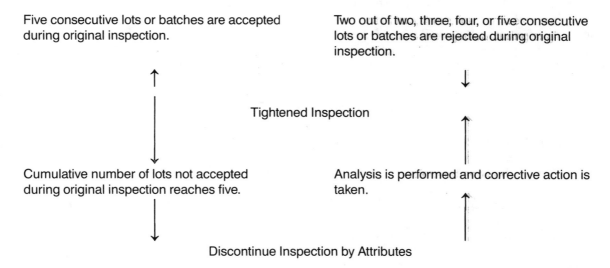

Normal Inspection

Five consecutive lots or batches are accepted during original inspection.

Two out of two, three, four, or five consecutive lots or batches are rejected during original inspection.

Tightened Inspection

Cumulative number of lots not accepted during original inspection reaches five.

Analysis is performed and corrective action is taken.

Discontinue Inspection by Attributes

FIGURE 2.6

Average Sample Size Curves

Table IX of the standard compares the average sample size of double and multiple sampling plans with a single sampling plan. These curves can be used with the information about the process to determine which approach to use. The process average, the AQL, and the acceptance number are used to determine the efficiency of using a double or multiple sampling plan versus a single sampling plan. For an AQL of 1 percent single-normal and an accept number of two, double inspection has an average sample size equal to that of single inspection when the process average is between 1 percent and 2 percent defective. In this case, there is no advantage in using a double inspection plan. Generally speaking, the administration and record keeping for double and multiple inspection are more difficult than for single inspection.

Differences Between 105D and 105E

Revision E of the standard is considerably different than revision D. It is formatted differently. The requirements and several of the definitions have been modified. All of the defined terms in revision E are included in one section, titled "Definitions." The definitions for the terms inspection, inspection by attributes, unit of product, major defect, minor defect, lot or batch size, defects per hundred units, percent defective, sample, and sample plan are the same in revision D and E.

In revision E, the definitions for critical defect, critical defective, major defective, minor defective, lot or batch, average outgoing quality (AOQ), and average outgoing quality limit (AOQL) were modified slightly from revision D.

The definitions for the terms defect, process average, sample size code letter, classification of defects, and acceptable quality level in revision E are either new or have been modified significantly from revision D.

The modified definition of acceptable quality level emphasizes that it is applicable to a continuous series of lots and that the process average is known and satisfactory. This definition carries with it the interpretation that a process capability study should be conducted, that the process should be in control, and that the process average should be considered satisfactory. The standard does not state who should determine whether the process average is satisfactory. Also new to this definition is a long note that points out lots or batches are more often accepted when the percent defective is no greater than the AQL.

Revision E includes a statement that, when MIL-STD-105E is referenced in a procedure, the procedure shall comply with the standard and reference appropriate sections. If a written procedure is prepared, it should be made available, upon request, to a government representative for review.

Steps to Obtaining a Sampling Plan

To determine which sampling plan to use, follow these steps:

1. Determine the lot size.

2. Determine the inspection level (inspection level I, II, or III or special level S-1, S-2, S-3, or S-4).

3. Locate where the row of lot sizes intersects the column for the selected inspection level.

4. Read the inspection code letter.

5. Select an AQL consistent with the process capability.

6. Determine whether single, double, or multiple inspection is needed.

7. Determine whether normal, tightened, or reduced inspection is needed.

8. Locate the correct table for the types of inspection selected in steps 6 and 7.

9. Using the inspection code letter determined in step 4, locate where the row for that code letter intersects the column for the selected AQL.

10. If there are values in the intersection, the sample size is shown in the column labeled "Sample Size." The acceptance number and reject number are provided in the "Ac" and "Re" columns, respectively.

11. If there is an arrow instead of accept and reject values, there is no sample plan for that code letter and AQL. Follow the arrow until Ac and Re values or other instructions are given.

12. If values are encountered, the new sample size is found where the row formed by the first set of accept/reject numbers intersects the sample size column.

13. If other instructions are provided, such as use corresponding single sampling plan, follow them.

3
DEMONSTRATING ACCEPTANCE SAMPLING BY ATTRIBUTES PER MIL-STD-105E

The following provides a visual method for demonstrating sampling inspection. It can be used to train inspectors or demonstrate to management the risks being taken when sampling inspection is used. The demonstration follows:

1. Invite 10 to 15 individuals to the demonstration. The demonstration will work with fewer individuals, however, it is more effective using a larger group.

2. Distribute a "lot" of material to each person. This lot, Figure 3.1, consists of one or more sheets of paper having a grid drawn on them. For the first demonstration, use one sheet of paper for the lot. Each box of the grid is a part.

3. Ask that each individual count the number of parts in the lot and write that number down. Even in this exercise, individuals might make a mistake in counting the lot. (If one sheet is used, the lot size is 304.) Discuss the ramification of a count error, whether a mistake is made or not.

4. As a group, select a sample plan, such as 4 percent general inspection level II, from MIL-STD-105E. The sample plan is selected using Table I—Sample Size Code Letters on page 13 of MIL-STD-105E (page 9 of revision D). Point out the lot or batch size is given on the left. Once the lot size is known, the lot is contained in only one given category. The lot size of 304 fits in the 281-to-500 category. Observing the top headings, find the columns headed by "General Inspection Levels." In this group, inspection levels I, II, and III are given. Since the chosen plan is general inspection level II, follow the column headed by "II" down until it intersects the line with the 281-to-500 batch size. Letter H is given as the code letter. The next step is to determine which inspection plan table to use.

   Page 14 of the standard contains Table II-A—Single Sampling Plans for Normal Inspection (Master Table). The code letters are given on the left. Follow this column down until the letter H is found. Locate the intersection of this row with the column headed by "4.0." The number five appears in the Ac column, the number six appears in the Re column, and the number 50 appears in the sample size column. Therefore, the sampling plan for a 4 percent general inspection level II has a sample size of 50. The lot is accepted if five or fewer defects are found and is rejected if six or more defects are found.

5. After determining the sampling plan, have each person put a dark X in 50 of the boxes, representing the sample. Measure the time taken for this activity. This time represents how long it takes to just pull the sample. Most likely, the time to pick this sample will be much shorter than the time it takes to pull a sample from a real lot since this example inspection activity does not involve material movement. The entire lot is immediately accessible and the parts do not need to be unpackaged. In many real inspection activities, a random sample is very difficult to take because of the product size or accessibility or the nature of the batch.

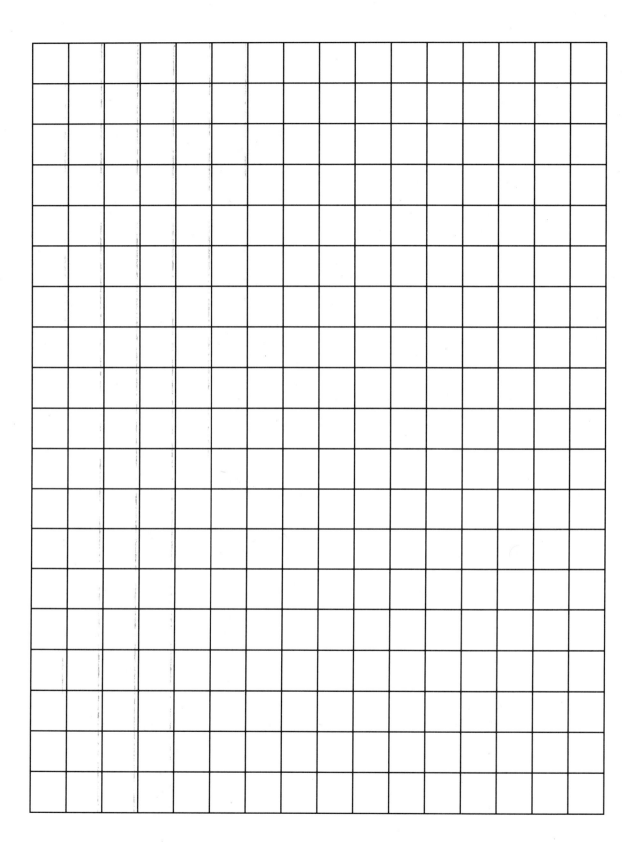

FIGURE   3.1

6. Some individuals might select a sample by marking out all the boxes in one particular area (for example, an area five boxes wide by 10 boxes deep). If this happens, use those samples to discuss the requirement and logic behind a random sample.

7. After each individual has selected his or her sample, use transparency 1 (Figure 3.2). Explain that the colored boxes on the transparency are defective parts. Point out that these 12 defects result in a nearly 4 percent defective rate (12/304 = 3.95 percent). Note transparency 1 has all the defects in one corner. Use this example to discuss the implications.

Further explain the concept that sampling inspection assumes defects are randomly distributed. Point out that this lot would be rejected based on pure luck, not statistical evidence. After reviewing all the aspects of nonrandom location of the defects, lay the transparency on each individual's marked sheet. Wherever an X and a colored box coincide, a defect is found. For anyone detecting six or more defects, the lot is rejected. For anyone detecting five or fewer defects, the lot is accepted. Recall that a 4 percent AQL general inspection level II is being used and the lot in question will only be rejected by luck.

8. After finishing with transparency 1, use transparency 2 (Figure 3.3). This sheet also has 12 defects placed in random locations. The lot is approximately 4 percent defective. Lay this transparency on each individual's marked sheet. Again, wherever an X and a colored box coincide, a defect is found. The lot is rejected if six or more defects are discovered; the lot is accepted if five or fewer defects are found. When comparing the transparency with the marked sheets, point out that another lot of 4 percent defective material was accepted. Statistically speaking, approximately 97 of 100 lots will be put into stock with the 4 percent defectives included. As a result, it is very unlikely that a lot will be rejected.

9. After finishing with transparency 2, use transparency 3 (Figure 3.4). This transparency has 24 defects, equivalent to approximately an 8 percent defective rate. The marked boxes are randomly located. Again, lay the transparency over the sheets marked with the X's. Wherever an X and a colored box coincide, a defect is found. Finding six or more defects rejects the lot, otherwise the lot is accepted. Note aloud each time a lot of 8 percent defective material is accepted. Statistically speaking, if switching procedures are not used, approximately 80 of 100 lots will be accepted with 8 percent defectives.

In actual use, 80 of 100 lots would be not accepted if MIL-STD-105E is used correctly. A tightened inspection plan would be used when two of five consecutive lots are rejected. Step 11 demonstrates the switching techniques.

10. Using transparency 4 (Figure 3.5)—which has 36 defects, making the lot approximately 12 percent defective—repeat the steps above. In this case, some individuals still accept the lot.

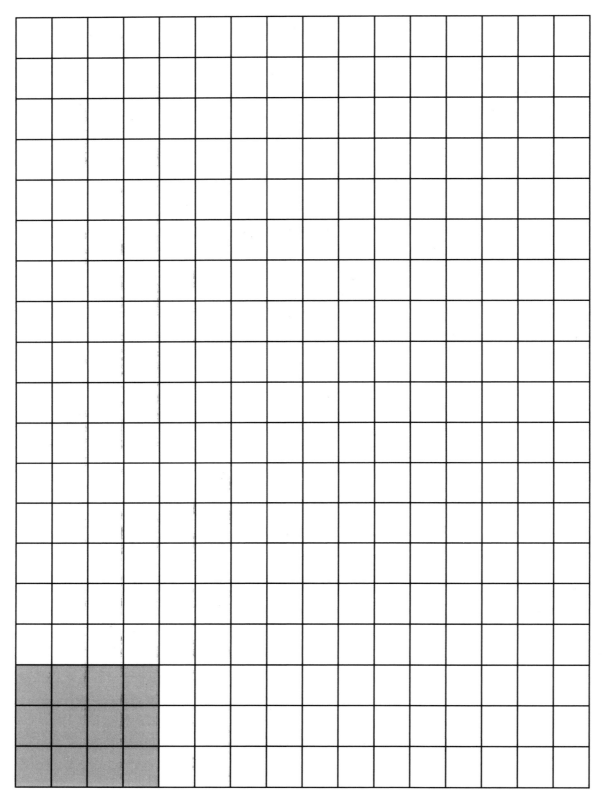

Transparency 1

FIGURE   3.2

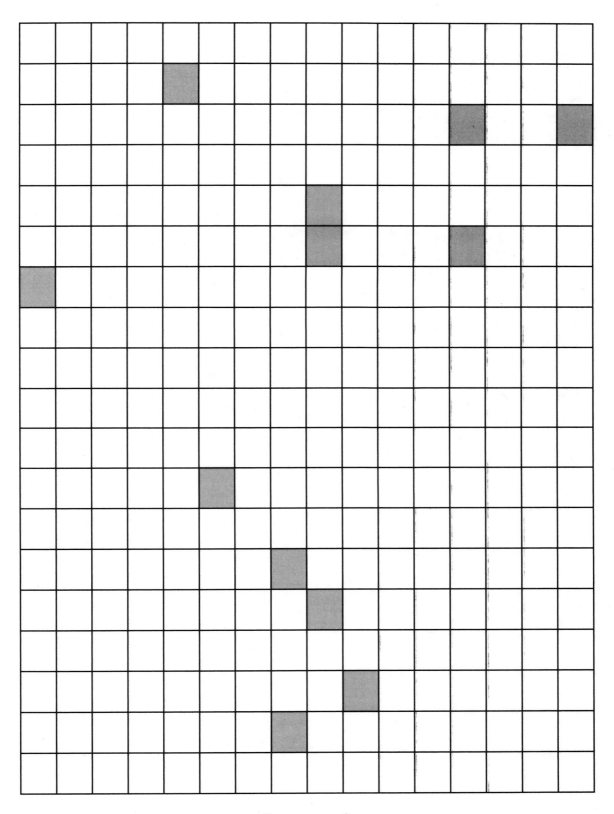

Transparency 2

FIGURE   3.3

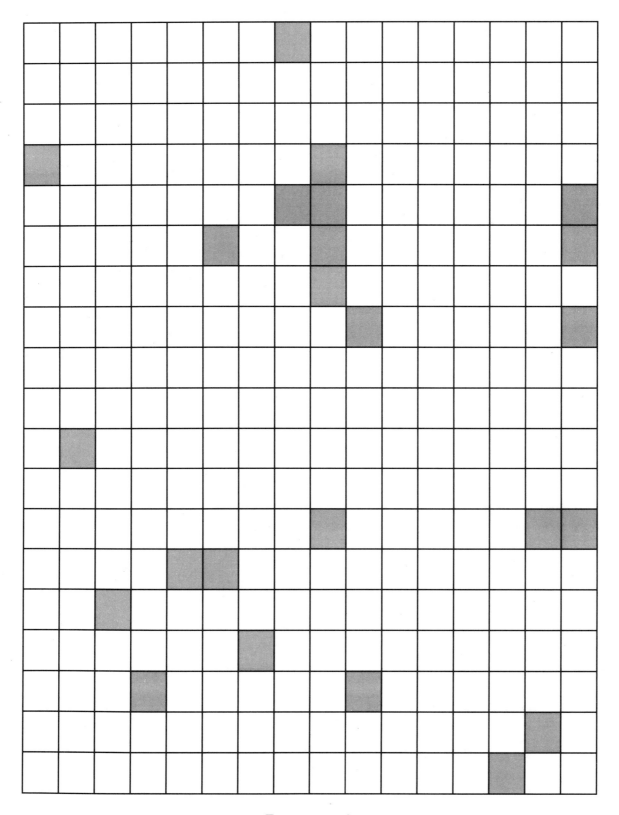

Transparency 3

FIGURE   3.4

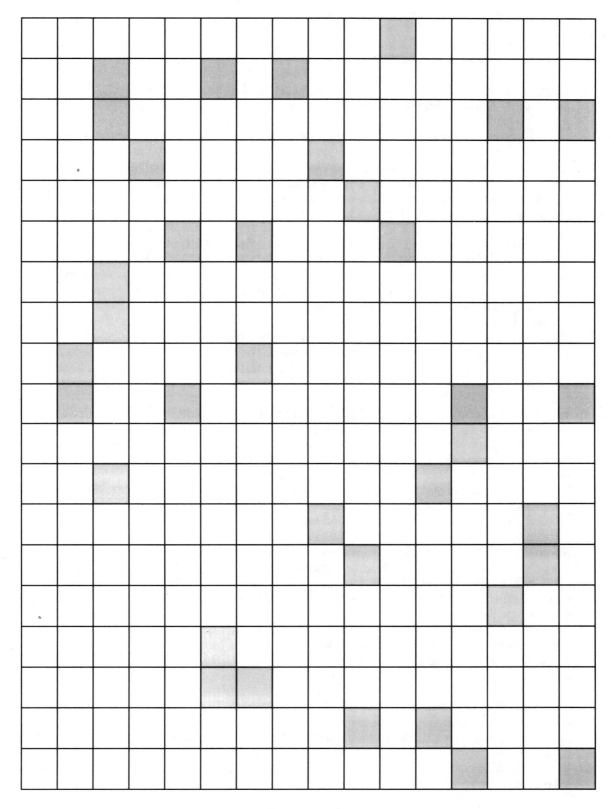

Transparency 4

FIGURE 3.5

At some point during this demonstration, make these observations:

- A significant amount of inspection effort is being expended, yet defective material is accepted and placed in stock or passed to the next process.

- In the example, 50 parts were handled and evaluated, and all the appropriate documentation was completed accurately. Still, defective material was placed in stock or passed to the next process.

- When this material is placed into the manufacturing cycle, it affects the total business unit.

- Material systems designed to minimize inventory assume all parts delivered to stock are good. In some cases, the material system incorporates factors to compensate for a level of loss during the process. However, this doesn't achieve the least inventory and results in manpower and machinery being used to produce unneeded parts.

11. Demonstrate how to switch from normal to tightened inspection and from normal to reduced inspection. Use one sheet of 304 parts as the lot. First, determine the sampling plans for tightened and reduced inspections. In both cases, start the process with inspection code letter H, as determined in step 4. Use Table II-B—Single Sampling Plans for Tightened Inspection (Master Table) from page 15 of MIL-STD-105E. On the left side of the table, find the column titled "Sample Size Code Letter" and follow it down to the letter H. Locate where the row with the H intersects the column headed by "4.0." The number three appears in the Ac column, the number four appears in the Re column, and the number 50 appears in the sample size column. Therefore, the tightened inspection plan uses a sample size of 50. The lot is accepted if three or fewer defects are found and is rejected if four or more defects are found.

The reduced inspection plan can be determined in a similar fashion using Table II-C—Single Sampling Plans for Reduced Inspection (Master Table) on page 16 of the standard. Find the column labeled "Sample Size Code Letter" and follow it down until the letter H is found. Locate where the row with the H intersects the column headed by "4.0." The number two appears in the Ac column, the number five appears in the Re column, and the number 20 appears in the sample size column. Therefore, the reduced inspection plan uses a sample size of 20. The lot is accepted if two or fewer defects are found and is rejected if five or more defects are found. If three or four defects are found, the lot is accepted, however, normal inspection is reinstated on the next lot.

Now that the sampling plans have been determined, return to transparency 3. Again compare the transparency with the marked sheets while using the normal inspection plan (sample size of 50, accept on five defects, and reject on six defects). As soon as the first lot is rejected, begin counting the lots inspected. The first lot rejected is number one, the next lot inspected, whether it is rejected or not, is number two, and so on until a total of five lots are counted. If no additional rejections occurred, explain that the switching procedure was not invoked. If another rejection occurs at lot number two, three, four, or five, the next lot is inspected with the tightened plan. This lot is inspected using the sample size of 50 and rejected if four or more defects are found. It is accepted if three or fewer defects are found.

Continue the comparisons using the tightened criteria, counting the number of lots being inspected with this plan. Point out that as soon as five lots are rejected, MIL-STD-105E paragraph 4.8 requires that sampling inspection be discontinued because the process shifted.

12. Demonstrate how to switch from tightened to normal inspection using the tightened inspection criteria and transparency 5 (Figure 3.6). The lot defective level for this case is 1 percent. Continue to compare lots until five consecutive lots are accepted. When this occurs, normal inspection is reinstated on the next lot.

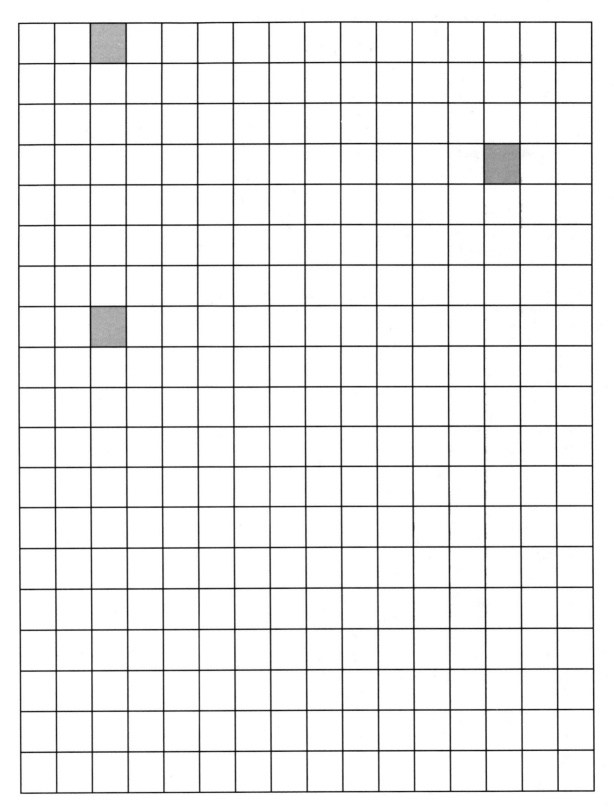

Transparency 5

FIGURE   3.6

13. To demonstrate how to switch from normal to reduced inspection, repeat the demonstration process using the criteria for a 4 percent AQL level II (inspect 50 parts, accept the lot if five defects are found, and reject the lot if six defects are found). Use transparency 5. Record the number of defects found in each inspection. If 10 consecutive lots are accepted, total the number of defects found during the inspection of the 10 lots. If the quantity is equal to or less than 14, reduced inspection can be used; if not, normal inspection continues.

14. Once reduced inspection is started, use transparency 3. Recall that transparency 3 has an 8 percent defective rate, uses 4 percent AQL reduced inspection, uses a sample of 20, accepts lots with two or fewer defects, and rejects lots with five or more defects. If three or four defects are found, the lot is accepted, however, normal inspection is reinstated.

    To demonstrate reduced inspection and how to switch to normal inspection, a new lot sheet has to be distributed. Each individual needs to randomly select a sample size of 20. Compare transparency 3 with the inspected lots applying the aforementioned criteria. Continue to compare the transparency with the inspected lots until normal inspection is required. Note that once normal inspection is restarted all the criteria in step 13 must be met to return to reduced inspection.

15. Demonstrate how to distribute a sample when the sublots are different quantities. Also demonstrate what should be done when no sample plan is given in the standard table.

    Distribute three sheets of lined paper to the group. Explain that these represent two subgroups. One subgroup is represented by one page of 304 parts, and the second subgroup is represented by 608 parts. The combination of the subgroups represents the total lot of 912 parts. Point out that paragraph 4.5.1 of the standard requires that the number of units in the sample be selected in proportion to the size of the sublots.

    For variety, select general inspection level III with an AQL of 0.04 percent. Using Table I, determine the inspection code letter. The quantity 912 falls within the 501-to-1,200 lot size. Find where the row for 501 to 1,200 intersects the column headed "General Inspection Level III." The letter K appears in this position. Using Table II, find where the row with the letter K intersects the column headed by "0.04."

    In this intersection, no Ac or Re number is given. In accordance with paragraph 4.9.3 of the standard, follow the arrow in the 0.04 column until Ac and Re numbers are given. This now coincides with a sample size of 315, but there is no sampling plan for the code letter K and an AQL of 0.04 percent. The first sample plan corresponding with an AQL of 0.04 is code letter M with a sample size of 315, an Ac of zero, and an Re of one. In the example, 105 samples are to be taken from the one lot represented by one sheet and 210 samples are to be taken from the two-sheet lot. The sample is distributed in proportion to the size of the sublots.

    If desired, have the group members mark their sheets with the desired random sample and compare them with transparencies 6A, 6B, and 6C, which have only two defects in three sheets (see Figures 3.7, 3.8, and 3.9). Measure the time it takes to do this keeping in mind that this only represents taking the sample from easily accessible lot and parts. The actual inspection will require its own time period.

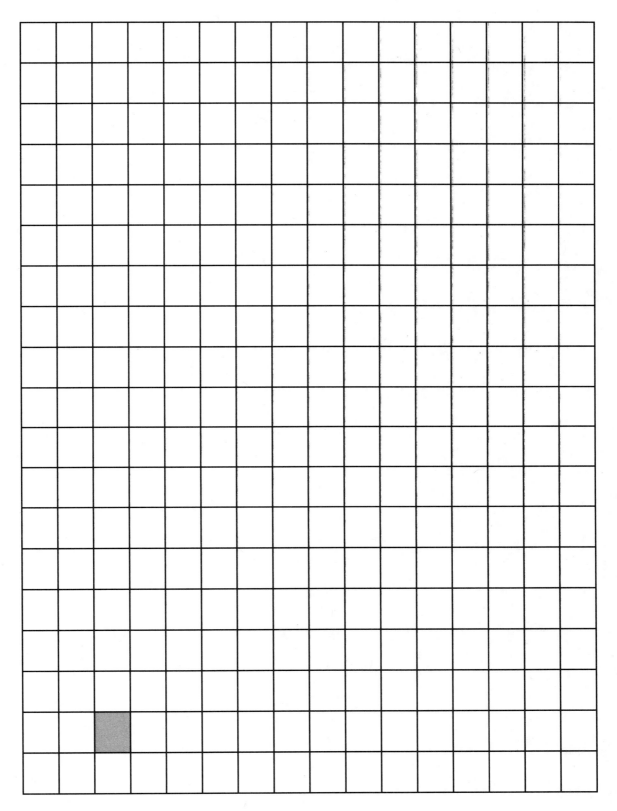

Transparency 6A

FIGURE   3.7

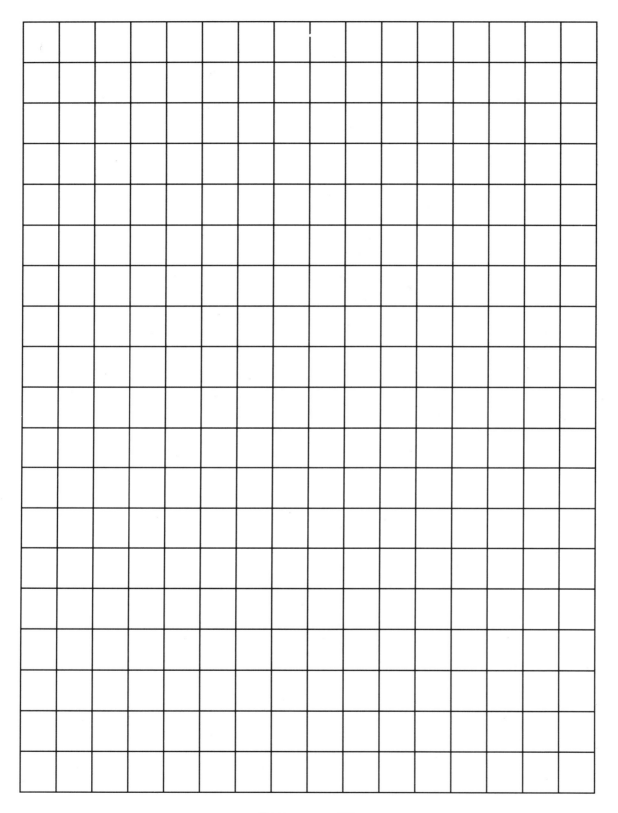

Transparency 6B

FIGURE   3.8

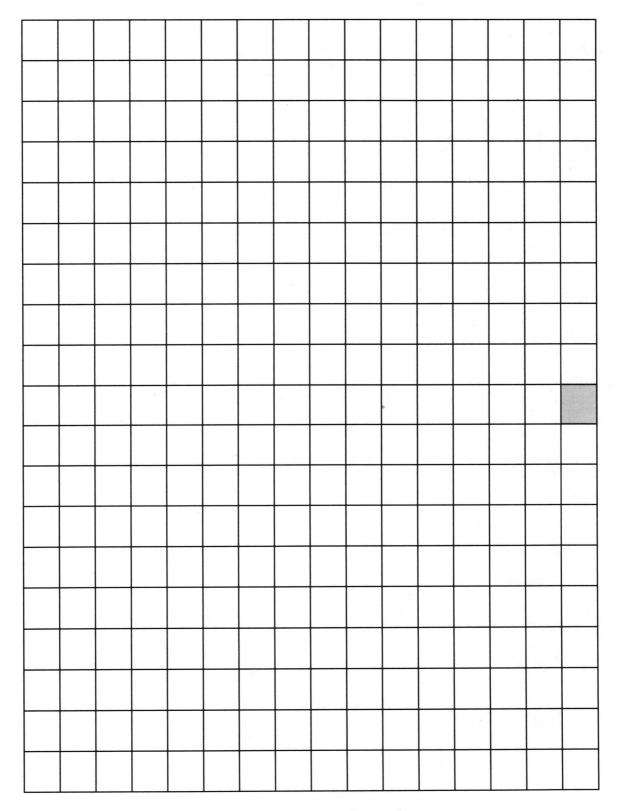

Transparency 6C

FIGURE 3.9

Double Inspection

Use the same approach to demonstrate double sampling inspection and other characteristics:

1. Distribute four sheets of ruled paper to each person. Have each person determine the number of parts included. (There are 1,208 parts.)

2. Choose an inspection level and AQL. For this example, special inspection level S-2 with an AQL of 1 percent is used. Recall that the special inspection levels are used when relatively small samples are needed and the associated high sampling risk can be tolerated.

3. Determine the inspection code letter by using Table I. Locate the lot or batch size that includes the lot in question. The 1,201-to-3,200 range includes the 1,208 count. Find the code letter at the intersection of the S-2 column and 1,201-to-3,200 row. This code letter is D.

4. Start with the normal inspection chart on page 17 of MIL-STD-105E, Table III-A — Double Sampling Plans for Normal Inspection (Master Table). Locate D in the column headed "Sample Size Code Letter." Find where the row with D intersects the column headed by "1.0." There is no acceptance or reject number given. Following the arrow down, an asterisk is encountered. The note at the bottom of the page states that, in this case, the single sampling plan is to be used. Explain that there is no double sample plan available for the code letter D and 1 percent AQL. The equivalent single sample plan is to be used. If desired, determine the equivalent sample plan using the technique described earlier.

5. Continue by choosing a different inspection level and AQL. For example, use general inspection level I and a 1 percent AQL. Starting again with Table I, determine the sample size code letter by identifying the letter at the intersection of the general inspection level I column and the 1,201-to-3,200 lot size row. It is inspection code letter H.

6. Referring to Table III-A, locate where the row with the code letter H and the column headed by "1.0" intersect. There are two sets of numbers. The first set has an Ac of zero and an Re of two. These correspond with a sample size of 32. This is the first sample criteria. The second set has an Ac of one and an Re of two. These also correspond to a sample size of 32. This is the second sample criteria.

7. Have each individual select the first sample of 32 from the entire lot. The sample must be distributed among the sublots and chosen randomly. Number the sublots and then number each part in the sublots. Using a random number table, the sample can be selected by using the first number of the random number to designate which sublot to take the sample from. For instance, 2,045 would designate part 45 in the second sublot.

8. Compare the four transparencies D1, D2, D3, and D4 (Figures 3.10, 3.11, 3.12, and 3.13) with an individual's inspection sample sheet. The transparencies have approximately a 1.5 percent defective rate, or 18 defects. If no defects are found, the lot is accepted. If two defects are found, the lot is rejected. If one defect is found, the lot is neither accepted nor rejected; instead, a second sample of 32 is selected.

   Another person's inspection sheet can be used for the second sample. It should be noted that, during actual inspection, the first sample should be returned to the lot before the second sample is taken so that those samples have equal chance of being selected. If during the second sample, no additional defects are found (i.e., the total number of defects from the first and second sample is one), the lot is accepted. If one or more defects are found in the second sample (i.e., the total number of defects from the first and second sample is two or more), the lot is rejected.

   Compare enough of the marked sheets with transparencies D1, D2, D3, and D4 until all possibilities occur.

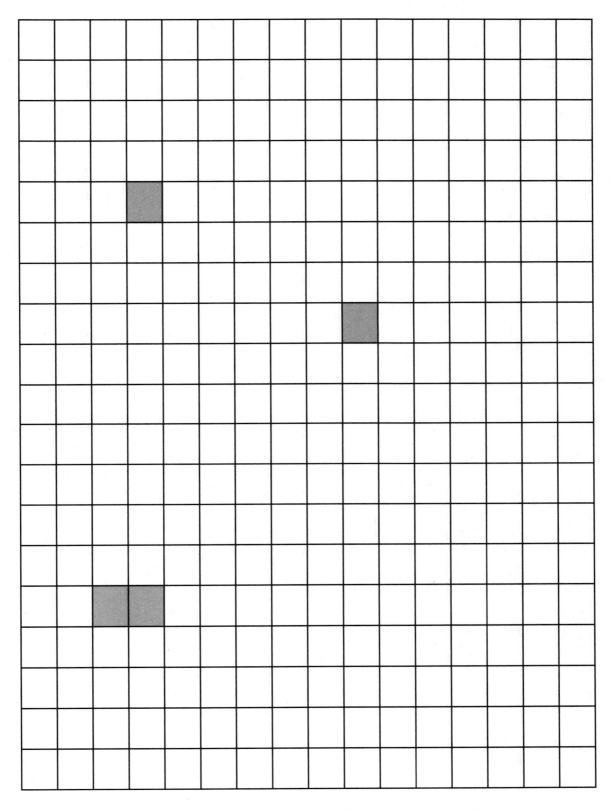

Transparency D1

FIGURE 3.10

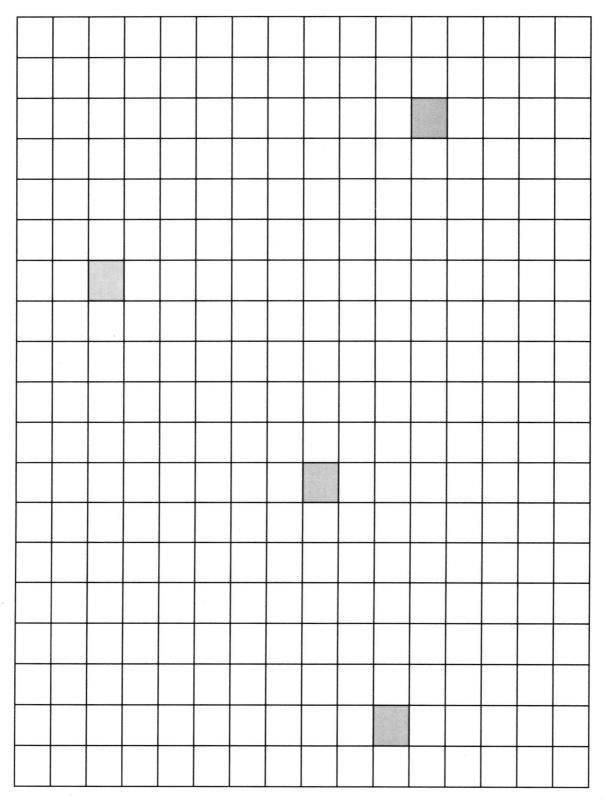

Transparency D2

FIGURE   3.11

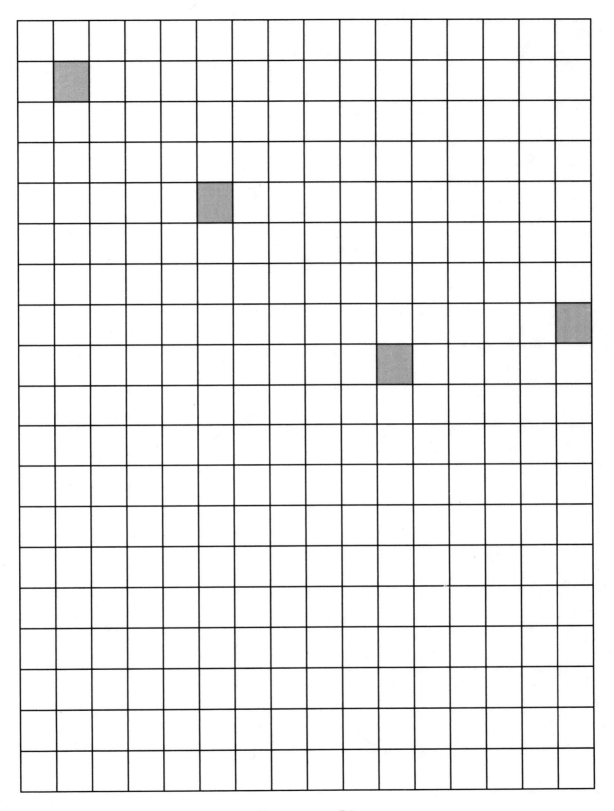

Transparency D3

FIGURE 3.12

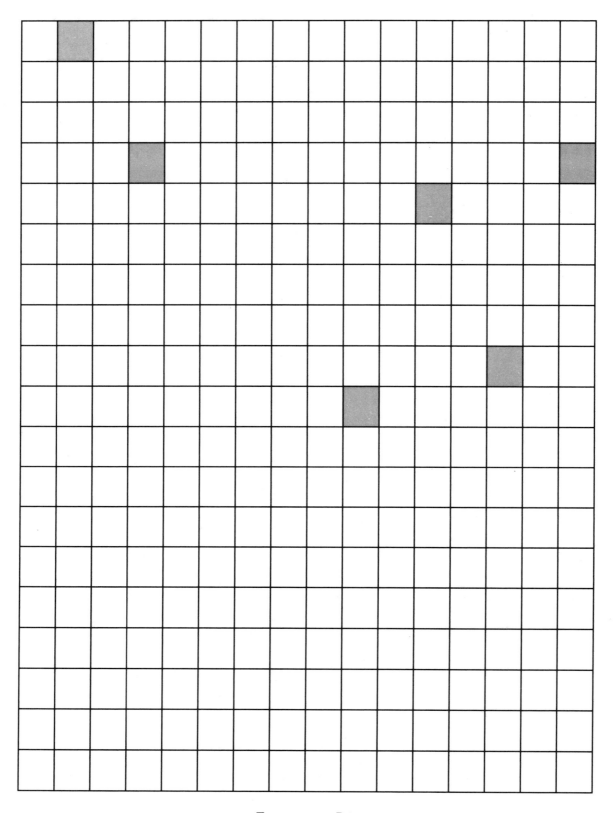

Transparency D4

FIGURE   3.13

9. Reduced and tightened inspection for double sample inspection is accomplished using the same procedures employed for single inspection, except Tables III-A, III-B, and III-C are used. If desired, demonstrate these procedures in a similar fashion.

Multiple Inspection

Using the same approach, demonstrate the multiple sampling approach:

1. Distribute one sheet of lined paper that has 304 parts. As in single sample inspection, start by choosing an inspection level. In this example, general inspection level III is used. Find where the row with 281 to 500 intersects the column labeled "III." The code letter J is given.

2. Choose an AQL. Recall that the AQL should be consistent with the process average. For this example, an AQL of 4 percent is used.

3. Using Table IV-A—Multiple Sampling Plans for Normal Inspection (Master Table) on page 20 of MIL-STD-105E and the intersection of 4, locate the sample size code letter J in the table. Verify that a sample plan exists in the 4 column. In this case, it does. The following plan appears:

	Sample	Cumulative Sample Size	Ac	Re
First	20	20	0	4
Second	20	40	1	6
Third	20	60	3	8
Fourth	20	80	5	10
Fifth	20	100	7	11
Sixth	20	120	10	12
Seventh	20	140	13	14

4. The first sample of 20 is chosen by marking a dark X in 20 squares on the lined paper.

5. Compare the sample lot with transparency M1 (Figure 3.14). This has a 5 percent defective rate. If no defects are found, the lot is accepted. If four defects are found, the lot is rejected. If one, two, or three defects are found, another sample of 20 is taken. The next person's sample can be used for this. Sampling continues using an additional sample size of 20 and the corresponding Ac and Re numbers. The number of defects is accumulated from the first inspection and each additional inspection. At each sample, a decision is made whether to accept the lot, reject the lot, or continue sampling until the lot is accepted or rejected in accordance with the criteria. For example, if during the third sample of 20, three defects were found from the first, second, and third inspection, the lot is accepted. If the accumulated number of defects is four, five, six, or seven, a fourth sample is taken. If the accumulated number of defects is eight, the lot is rejected.

6. Reduced and tightened inspection samples are determined using the same method employed for single sample inspection.

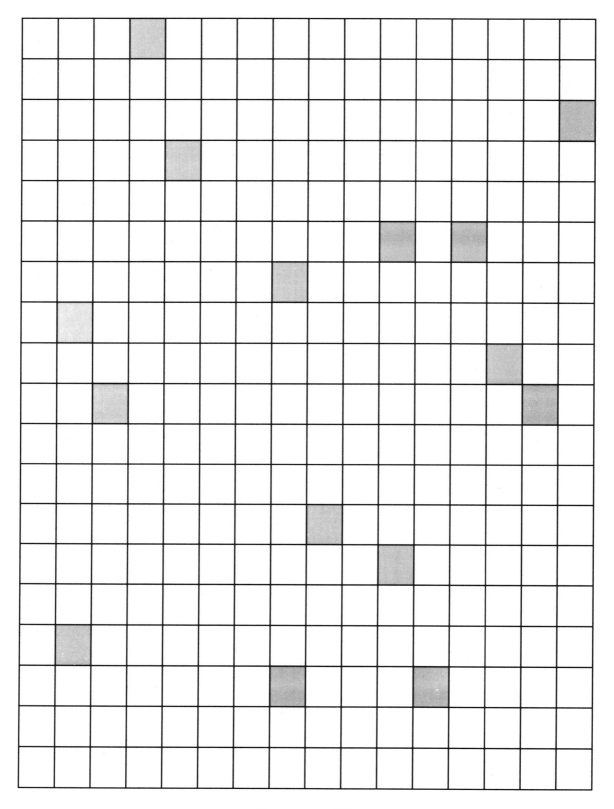

Transparency M1

FIGURE   3.14

Average Outgoing Quality (AOQ)

The AOQ can be demonstrated using the single sample approach with a lot having 5 percent defective. Compare transparency M1 with the sample. For each lot accepted, accumulate the number of good parts and bad parts. For each lot rejected, count only the good parts. The bad parts are removed through 100 percent inspection. The total of the bad parts divided by the total of the good parts is representative of the AOQ. Compare the resultant quotient with that shown in Table V-A on page 26 in the standard.

4

PROCEDURE FOR SAMPLING INSPECTION BY ATTRIBUTES; PREPARATION OF INSPECTION PLANNING AND INSPECTION RECORD

1.0 Purpose

1.1 The purpose of this procedure is to provide:

    1.1.1 The method and forms to be used for developing and preparing inspection plans.

    1.1.2 The definition of terms to be used during preparation, inspection, and other applicable activities.

    1.1.3 An inspection approach for evaluating products through sampling inspections, where all characteristics might not be measured on a day-by-day or lot-by-lot basis, but control is maintained.

    1.1.4 A procedure for implementing sampling inspection by attributes (see paragraph 4.1 in MIL-STD-105E for more information).

1.2 The applicable specifications sampling tables are:

    1.2.1 MIL-STD-105E, Inspection Sampling by Attributes

    1.2.2 MIL-STD-109, QA Terms and Definitions

    1.2.3 MIL-Q-9858, Quality Systems Requirements

2.0 Application

2.1 Sampling plans described in this procedure are applicable, but not limited to, inspection of the following:

    2.1.1 Incoming components and raw material

    2.1.2 Selected manufacturing and assembly operations

    2.1.3 Processes

    2.1.4 Documentation

    2.1.5 Operations of services

2.1.6 Administrative procedures

2.2 Unless specified in the quality manual, inspection instructions, or acceptance procedure, sampling inspection will not be used for final acceptance of completed end items. Customer specifications or contract requirements on sampling for final acceptance of completed items take precedence. Specific customer or contract requirements also take precedence.

2.3 The requirements of this instruction are applicable to inspection instructions prepared after (insert own date). Inspection instructions prepared prior to (insert own date) shall be in accordance with (insert own number).

2.4 The use of acceptance sampling by attributes does not imply that products are accepted if they meet the criteria of this procedure. Product acceptance criteria are defined in the customer contract (see paragraph 1.1 in MIL-STD-105E for more information).

3.0 Responsibility

3.1 The following persons have responsibilities that are described in this procedure:

3.1.1 Quality managers

3.1.2 Quality engineers

3.1.3 Quality analysts

3.1.4 QC inspection supervisors

3.1.5 Process control engineers

3.1.6 Inspection planners

3.1.7 Inspectors

3.1.8 (Insert own title)

4.0 Procedure

4.1 Preparation of inspection plans:

4.1.1 The following may designate individuals to prepare inspection plans:

4.1.1.1 Managers and supervisors in the quality organization

4.2 For the purpose of this instruction, the person preparing the inspection plan will be referred to as the inspection planner. The person's position or title need not be inspection planner.

4.3 Quality managers, quality supervisors, quality analysts, and quality engineers may prepare inspection plans.

4.4 The individuals identified in paragraph 4.1.1.1 shall maintain a record similar to Figure 4.1, which identifies the persons authorized to prepare plans. This record is evidence that the inspection planner completed his or her training in accordance with the requirements of this document.

4.5 Quality engineers will prepare the initial inspection plan for new products, their associated subassemblies, and detailed parts.

Training Record

Name of Individual   Date of Training   Department   Individual Accomplishing Training

Note: A form similar to this shall be maintained to identify persons authorized to prepare plans.

FIGURE   4.1

4.6 During the preparation of the initial inspection plan, quality engineers shall identify inspection requirements that are specified within customer documentation. In addition, quality engineers shall provide a method for updating appropriate plans as customer requirements change.

4.7 The inspection planner will determine inspection characteristics, characteristic classification, sample plan, and equipment for inspection by taking into consideration these items:

   4.7.1 The process variability as established by historical records or process capability studies.

   4.7.2 Print, process, and other specifications that are applicable.

   4.7.3 The method used to produce the part and variability of the process.

   4.7.4 The use of the part in the next application.

   4.7.5 The likelihood for a defect being found during assembly.

   4.7.6 The potential effect of a defect on the unit's operation or individual using it or maintaining it.

   4.7.7 The quantity of parts used in the next assembly.

   4.7.8 Historical information, such as inspection records, process capability studies, and first-article information.

   4.7.9 The amount of in-process or other inspection or verification applied to the part.

   4.7.10 Data from industry-established capabilities.

4.8 When generating or revising an inspection plan, the inspection planner shall use the terms and definitions provided in this procedure.

4.9 All inspection plans shall be reviewed and approved by the individuals listed in paragraph 4.1.1. Any quality engineer, quality analyst, or group leader for the area can also approve an inspection plan.

4.10 During review and approval, the factors noted in paragraph 4.7 shall be considered.

4.11 Inspection plans may be modified by individuals authorized to prepare plans. Modification will be based on print revisions, inspection findings (both good and bad), changes in the process, or customer problems where known.

   Inspection plan changes—i.e., additions, deletions, or correction or characteristics called out for inspection—will be signed and dated in the appropriate blocks by the designated approval personnel.

4.12 The following identifies the information to be filled in on the inspection plan shown in Figure 4.2. Areas 12 through 23 are to be entered during the inspection activity. The remainder is to be entered during plan preparation.

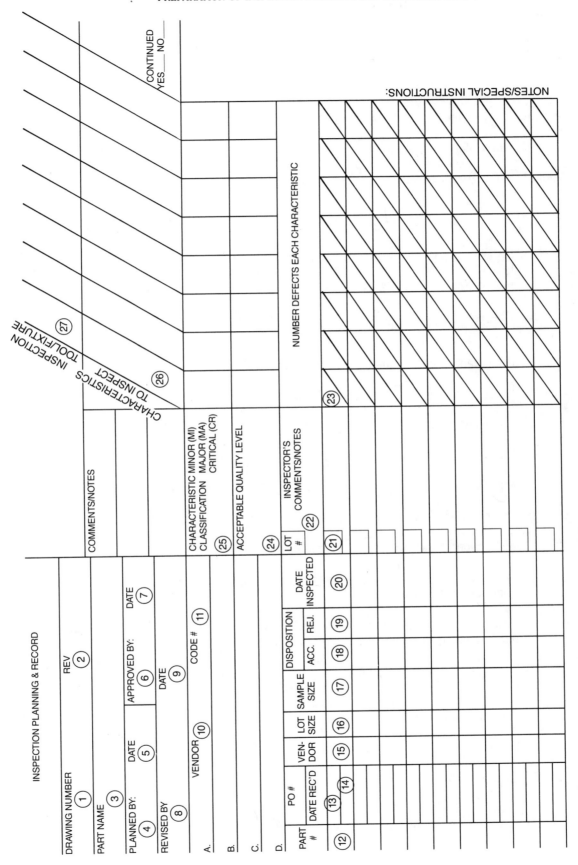

FIGURE 4.2

Area 1. Enter drawing number.

Area 2. Enter current print revision letter.

Area 3. Enter part name.

Area 4. Enter name of person originating the plan.

Area 5. Enter plan completion date.

Area 6. The manager, supervisor, or designated individual who receives and approves the plan signs here.

Area 7. Enter date the plan is approved.

Areas 8 & 9. Enter the name of the person performing a review or revision in event of a drawing or specification change and the date the review or revision was performed. This area will be left blank if no revision was made.

Area 10. Enter source name. If at receiving, enter supplier's name. Also list other known approved suppliers.

Area 11. Enter vendor code number, if applicable.

Area 12. Enter dash number of part, if applicable.

Area 13. Enter purchase order number or production release number.

Area 14. Enter date the current lot was submitted for inspection or received, as applicable.

Area 15. Enter A, B, C, or D as applicable for current vendor or source.

Area 16. Enter lot size. Note: Prior to inspection, the product shall be assembled into an identifiable lot, sublot, or batch. Each lot, as far as practical, shall consist of units of product of a single type, grade, class, size, or composition.

Area 17. Enter sample size. This number will reflect the largest number recorded in the lower half of any block in area 23.

Area 18. Enter total number of conforming parts. This number might not be the total number of parts accepted (i.e., some nonconforming parts might be accepted as a result of material review, etc.).

Area 19. Enter total number of nonconforming parts (see instructions for area 18). If the lot or the remainder of the lot is rejected based on sample inspection findings, this number will reflect the total quantity of parts submitted for rework, material review, return to supplier, etc.

Area 20. Enter date inspection is completed.

Area 21. Enter lot number, beginning with number one and using consecutively higher numbers as subsequent lots are inspected.

Area 22. Enter any comments pertinent to the inspection of disposition or the current lot.

Area 23. Lower half—Enter the applicable size on this particular characteristic based on the inspection level and AQL data given in area 24.

Upper half—Enter the total number of defective parts found in the sample lot, when applicable. If there were no defects, enter " — ".

Area 24. Enter the applicable inspection level and AQL determined to be adequate for the particular characteristic listed in area 26, considering part history and other pertinent factors.

Area 25. Enter the applicable designation MI, MA, CR, or X based on respective definitions outlined in paragraph 4.13.

Area 26. Enter the characteristic to be inspected, being specific when possible. List tolerances, description (body length, lead diameter, etc.), drawing zone location, etc.

Area 27. Enter inspection tool, inspection aid, holding fixture, or other equipment used to measure or help measure this particular characteristic.

Form XX (Figure 4.3) is a supplement that provides extended space to enter additional lot numbers and related data without repreparing the inspection plan.

4.13 Classification symbols and sampling plans

4.13.1 The following establishes and defines symbols used for classifying characteristics.

4.13.1.1 Critical (CR)—This symbol is used in the class inspection column of the inspection plan to denote critical defects and critical characteristics defined on drawings or from experience.

4.13.1.2 Major (MA)—This symbol is used in the class inspection column of the inspection plan to denote a major characteristic or defect.

4.13.1.3 Minor (MI)—This symbol is used in the class inspection column of the inspection plan to denote a minor characteristic or defect.

4.13.1.4 Other (X)—This symbol is used in the class inspection column to indicate a major characteristic for which 100 percent inspection is accomplished.

4.14 Selection of degree of inspection

4.14.1 Inspection planning, which uses sampling inspection, starts off under normal inspection and will stay under normal inspection until changed in accordance with the following conditions (see paragraph 4.6 in MIL-STD-105E for more information).

4.14.1.1 Normal to tightened—When normal inspection is in effect for a characteristic, tightened inspection shall be instituted when two of five consecutive lots or batches have been rejected for the characteristic during original inspection. Resubmitted lots or batches shall be ignored for this (see paragraph 4.7 in MIL-STD-105E for more information).

4.14.1.2 Tightened to normal—When tightened inspection is in effect for a characteristic, normal inspection shall be instituted when five consecutive lots or batches have been considered acceptable for the characteristic during original inspection. Note: This applies to five lots from the same process or supplier, as applicable (see paragraph 4.7.2 in MIL-STD-105E).

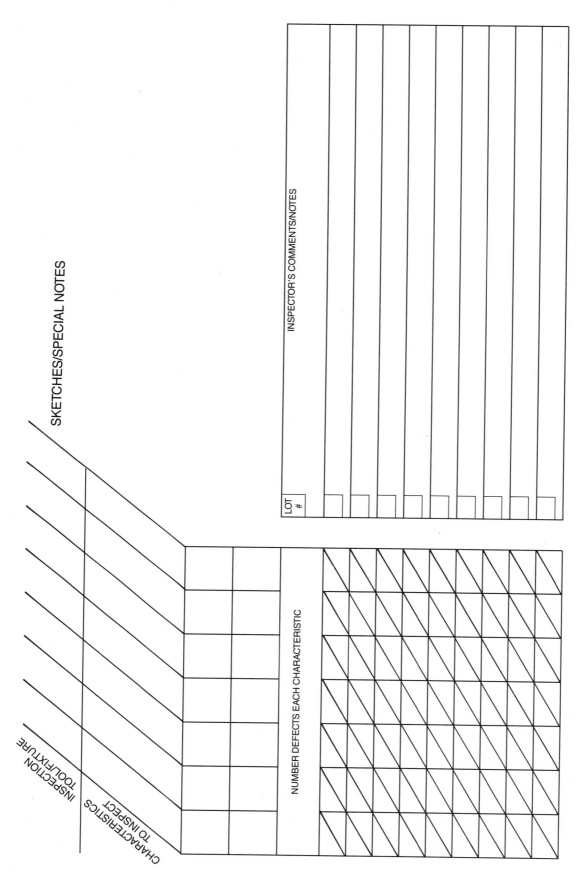

FIGURE 4.3

4.14.1.3 Normal to reduced—When normal inspection is in effect, reduced inspection may be instituted provided that all of the following conditions are satisfied (see paragraph 4.7.3 in MIL-STD-105E):

- The preceding 10 or more lots or batches have been on normal inspection and all have been accepted during original inspection.

- The total number of defects in the samples from the preceding 10 lots or batches is equal to or less than the applicable number given in Table VIII on page 32 of MIL-STD-105E. If double or multiple sampling is in use, all samples inspected should be included, not only first samples.

- Production is at a steady rate.

- Reduced inspection is approved by one of the individuals designated in paragraph 4.1.1.1.

4.14.1.4 Discontinuance of inspection — In the event that the cumulative number of lots not accepted in a sequence of consecutive lots on original tightened inspection reaches five, sampling inspection is to be discontinued. An investigation into the cause and appropriate corrective action are required prior to resumption of inspection sampling. When inspection sampling is resumed, it shall resume using the tightened inspection criteria (see paragraph 4.8 in MIL-STD-105E for more information).

5.0 Special Provisions

5.1 Training outline

5.1.1 Instruction for inspection planners shall cover as a minimum:

5.1.1.1 The need for uniform application of sampling methods.

5.1.1.2 The importance of record keeping.

5.1.1.3 The symbols for defects.

5.1.1.4 The methods for classifying characteristics and AQL.

5.1.1.5 How to select the degree of inspection.

5.1.1.6 Sampling procedures in MIL-STD-105E.

A Guideline for Inspection Sampling by Attributes

During inspection sampling by attributes, randomly selected units are used as samples and inspected to determine whether the characteristics of the items are within specification limits. The characteristics are considered to be good or bad, and each part is considered good or bad. Based on information generated from the inspection, a decision is made regarding the acceptability of the entire lot.

Each lot must be inspected in accordance with the instructions specified by the quality engineering department or in an inspection plan. If the results of the inspection for each characteristic are acceptable, the entire lot can be accepted. It is important to note that sampling inspection does not on its own determine acceptance of a lot of parts. Part and process conformance to specifications and contracts are the determining factors. If the number of parts found to have defects exceeds the accept limits for the sample size, the lot must be rejected and documented as such. The lot can be screened 100 percent for the defective characteristics. Only those parts found to be nonconforming need to be rejected.

The sample size must be in accordance with the appropriate sampling tables as specified by the designated level in the inspection plan.

A random sample must be drawn from each lot to be inspected. Sample pieces must be selected without regard to position or apparent quality. Each piece in the lot should have an equal chance of being selected. A random number table or other method should be used to ensure a random sample.

Inspection must be performed on all articles in the sample. All characteristics must be inspected in accordance with the instructions specified in the inspection plan.

In the event 100 percent checks are required for some characteristics, those characteristics requiring sample inspection must be selected and inspected first. If other defective characteristics are found during a 100 percent check of a particular characteristic, the part should be rejected even if that characteristic was not being inspected. The part is considered good or bad. It is good when all characteristics are acceptable. It is bad if one or more characteristics are defective.

Sample inspection should be performed using single, double, or multiple sampling inspection plans with an AQL specified by the inspection plan. The results of each characteristic inspected must be documented individually.

Example 1

You have 750 parts to inspect. Your inspection plan is general inspection level II with an AQL of 1 percent. You will perform normal inspection when there is no indication to change to tightened or reduced inspection. This is determined by studying the results of previous inspections. You verify that the part number and revision number match the part number and revision number of the requirements documentation. You observe characteristics that would suggest the parts were not manufactured from similar processes.

If a single sampling plan is used, the sample size is 80, the accept number is two, and the reject number is three. The entire lot will be accepted if two or fewer parts are found defective; the entire lot will be rejected if three or more parts are found defective. If the lot is accepted, the defective parts will be segregated from the lot and processed as nonconforming material. A rejected lot will be documented and processed as nonconforming material.

Example 2

Inspection of the same parts using the double sampling plan results in two sample sizes of 50 each. The first lot has an acceptance number of zero and a reject number of three. The second lot has an acceptance number of three and a reject number of four. The entire lot will be accepted if no defective parts are found and rejected if three defective parts are found during the first sample. If one or two defective parts are found during the first sample, a second sample needs to be inspected. If the second sample is used, lot acceptance or rejection is based on the total number of defects found during the first and second samples. The entire lot will be accepted if three or fewer parts from both samples are found defective. The entire lot will be rejected if a total of four or more defective parts are found from both samples.

5

DEFINITIONS

The following definitions are to be used during implementation of the procedure titled "Procedure for Sampling Inspection Attributes; Preparation of Inspection Planning and Inspection Record." The definitions contained herein match those of paragraph 3 (DEFINITIONS) of MIL-STD-105E.

Acceptable Quality Level (AQL). When a continuous series of lots is considered, the AQL is the quality level which, for the purposes of sampling inspection, is the limit of a satisfactory process average. (Reference 3.19 of MIL-STD-105E.)

Average Outgoing Quality (AOQ). For a particular process average, the AOQ is the average quality of outgoing product including all accepted lots or batches, plus all rejected lots or batches after the rejected lots or batches have been effectively 100 percent inspected and all defectives replaced by nondefectives.

Average Outgoing Quality Limit (AOQL). The AOQL is the maximum AOQ for a given acceptance sampling plan. Factors for computing AOQL values are given in Table V-A of MIL-STD-105E for each of the single sampling plans for normal inspection and in Table V-B of MIL-STD-105E for each of the single sampling plans for tightened inspection.

Classification of Defects. A classification of defects is the enumeration of possible defects of the unit of product classified according to their seriousness.

Critical Defect. A critical defect is a defect that judgement and experience indicate would result in hazardous or unsafe conditions for individuals using, maintaining, or depending upon the product, or a defect that judgement and experience indicate is likely to prevent performance of the tactical function of a major end item such as a ship, aircraft, tank, missile, or space vehicle.

Critical Defective. A critical defective is a unit of product which contains one or more critical defects and may also contain major and/or minor defects.

Defect. A defect is any nonconformance of the unit of product with specified requirements.

Defective. A defective is a unit of product which contains one or more defects.

Defects per Hundred Units. The number of defects per hundred units of any given quantity of units of product is 100 times the number of defects contained therein (one or more defects being possible in any unit of product) divided by the total number of units of product, i.e.:

Defects per hundred units = Number of defects $\times$ 100 / Number of units inspected

Inspection. Inspection is the process of measuring, examining, testing, or otherwise comparing the unit of product with the requirements.

Inspection by Attributes. Inspection by attributes is inspection whereby either the unit of product is classified simply as defective or nondefective or the number of defects in the unit of product is counted, with respect to a given requirement or set of requirements.

Lot or Batch. The term "lot" or "batch" shall mean "inspection lot" or "inspection batch," i.e., a collection of units of product from which a sample is to be drawn and inspected and may differ from a collection of units designated as a lot or batch for other purposes (e.g., production, shipment, etc.).

Lot or Batch Size. The lot or batch size is the number of units of product in a lot or batch.

Major Defect. A major defect is a defect, other than critical, that is likely to result in failure or to reduce materially the usability of the unit of product for its intended purpose.

Major Defective. A major defective is a unit of product which contains one or more major defects and may also contain minor defects, but contains no critical defect.

Minor defect. A minor defect is a defect that is not likely to reduce materially the usability of the unit of product for its intended purpose or is a departure from established standards having little bearing on the effective use or operation of the unit.

Minor Defective. A minor defective is a unit of product which contains one or more minor defects, but no critical or major defect.

Percent Defective. The percent defective of any given quantity of units of product is 100 times the number of defective units of product contained therein divided by the total number of units of product, i.e.;

$$\text{Percent defective} = \text{Number of defectives} \times 100 \,/\, \text{Number of units inspected}$$

Process Average. The process average is the average percent defective or average number of defects per hundred units (whichever is applicable) of product submitted by the supplier for original inspection. Original inspection is the first inspection of a particular quantity of product as distinguished from the inspection of product which has been resubmitted after prior rejection.

Sample. A sample consists of one or more units of product drawn from a lot or batch, the units of the sample being selected at random without regard to their quality. The number of units of product in the sample is the sample size.

Sample Size Code Letter. The sample size code letter is a device used along with the AQL for locating a sampling plan on a table of sampling plans.

Sampling Plan. A sampling plan indicates the number of units of product from each lot or batch which are to be inspected (sample size or series of sample sizes) and the criteria for determining the acceptability of the lot or batch (acceptance and rejection numbers).

Unit of Product. The unit of product is the thing inspected in order to determine its classification as defective or nondefective or to count the number of defects. It may be a single article, a pair, a set, a length, an area, an operation, a volume, a component of an end product, or the end product itself. The unit of product may or may not be the same as the unit of purchase, supply, production, or shipment.

6

TABLES FOR INSPECTION

Many industries develop tables showing sample size, accept numbers, and reject numbers. This is done to minimize the time and the potential for error in determining a sampling plan. Tables are effective when set sampling plans are used. This section includes several examples of the types of tables that can be given to inspectors. The tables are for:

- 1 percent   AQL single level II
- 4 percent   AQL single level II
- 0.65 percent AQL single level S-3
- 1 percent   AQL single level S-3
- 2.5 percent AQL single level S-3
- 1 percent   AQL double level II
- 4 percent   AQL double level II
- 1 percent   AQL double level S-3
- 2.5 percent AQL double level S-3

It must be noted that for double inspection, if the accept number is exceeded but the reject number is not reached, a second sample must be taken. Lot acceptance is then determined from the total number of defects found during the first and second samples.

Single Sampling Plan for 1 Percent AQL Level II Normal

Lot size	Sample size	Accept	Reject
2-8	100 percent	N/A	N/A
9-15	100 percent	N/A	N/A
16-25	13	0	1
26-50	13	0	1
51-90	13	0	1
91-150	13	0	1
151-280	50	1	2
281-500	50	1	2
501-1,200	80	2	3
1,201-3,200	125	3	4
3,201-10,000	200	5	6
10,001-35,000	315	7	8
35,001-150,000	500	10	11
150,001-500,000	800	14	15

Single Sampling Plan for 1 Percent AQL Level II Tightened

Lot size	Sample size	Accept	Reject
2-8	100 percent	N/A	N/A
9-15	100 percent	N/A	N/A
16-25	20	0	1
26-50	20	0	1
51-90	20	0	1
91-150	20	0	1
151-280	80	1	2
281-500	80	1	2
501-1,200	80	1	2
1,201-3,200	125	2	3
3,201-10,000	200	3	4
10,001-35,000	315	5	6
35,001-150,000	500	8	9
150,001-500,000	800	12	13

Single Sampling Plan for 1 Percent AQL Level II Reduced

Lot size	Sample size	Accept	Reject
2-8	5	0	1
9-15	5	0	1
16-25	5	0	1
26-50	5	0	1
51-90	5	0	1
91-150	5	0	1
151-280	20	0*	2
281-500	20	0	2
501-1,200	32	1	3
1,201-3,200	50	1	4
3,201-10,000	80	2	5
10,001-35,000	125	3	6
35,001-150,000	200	5	8
150,001-500,000	315	7	10

*If the number of defects exceeds the accept number but is less than the reject number, accept the lot and return to normal inspection.

Single Sampling Plan for 4 Percent AQL Level II Normal

Lot size	Sample size	Accept	Reject
2-8	3	0	1
9-15	3	0	1
16-25	3	0	1
26-50	13	1	3
51-90	13	1	3
91-150	20	2	3
151-280	32	3	4
281-500	50	5	6
501-1,200	80	7	8
1,201-3,200	125	10	11
3,201-10,000	200	14	15
10,001-35,000	315	21	22
35,001-150,000	315	21	22
150,001-500,000	315	21	22

Single Sampling Plan for 4 Percent AQL Level II Tightened

Lot size	Sample size	Accept	Reject
2-8	5	0	1
9-15	5	0	1
16-25	5	0	1
26-50	20	1	2
51-90	20	1	2
91-150	20	1	2
151-280	32	2	3
281-500	50	3	4
501-1,200	80	5	6
1,201-3,200	125	8	9
3,201-10,000	200	12	13
10,001-35,000	315	18	19
35,001-150,000	315	18	19
150,001-500,000	315	18	19

Single Sampling Plan for 4 Percent AQL Level II Reduced

Lot size	Sample size	Accept	Reject
2-8	2	0	1
9-15	2	0	1
16-25	2	0	1
26-50	5	0*	2
51-90	5	0	2
91-150	8	1	3
151-280	13	1	4
281-500	20	2	5
501-1,200	32	3	6
1,201-3,200	50	5	8
3,201-10,000	80	7	10
10,001-35,000	125	10	13
35,001-150,000	125	10	13
150,001-500,000	125	10	13

*If the number of defects exceeds the accept number but is less than the reject number, accept the lot and return to normal inspection.

Single Sampling Plan for 0.65 Percent AQL Level S-3 Normal

Lot size	Sample size	Accept	Reject
2-8	20	0	1
9-15			
16-25			
26-50			
51-90			
91-150			
151-280			
281-500			
501-1,200			
1,201-3,200			
3,201-10,000			
10,001-35,000			
35,001-150,000			
150,001-500,000	20	0	1

Single Sampling Plan for 0.65 Percent AQL Level S-3 Tightened

Lot size	Sample size	Accept	Reject
2-8	32	0	1
9-15			
16-25			
26-50			
51-90			
91-150			
151-280			
281-500			
501-1,200			
1,201-3,200			
3,201-10,000			
10,001-35,000			
35,001-150,000			
150,001-500,000	32	0	1

Single Sampling Plan for 0.65 Percent AQL Level S-3 Reduced

Lot size	Sample size	Accept	Reject
2-8	13	0	1
9-15			
16-25			
26-50			
51-90			
91-150			
151-280			
281-500			
501-1,200			
1,201-3,200			
3,201-10,000			
10,001-35,000			
35,001-150,000			
150,001-500,000	13	0	1

Note: For this level and AQL, the single sampling plan is to be used for the double sampling plan for all lot sizes.

Single Sampling Plan for 1 Percent AQL Level S-3 Normal

Lot size	Sample size	Accept	Reject
2-8	13	0	1
9-15			
16-25			
26-50			
51-90			
91-150			
151-280			
281-500			
501-1,200			
1,201-3,200			
3,201-10,000			
10,001-35,000	13	0	1
35,001-150,000	50	1	2
150,001-500,000	50	1	2

Single Sampling Plan for 1 Percent AQL Level S-3 Tightened

Lot size	Sample size	Accept	Reject
2-8	20	0	1
9-15			
16-25			
26-50			
51-90			
91-150			
151-280			
281-500			
501-1,200			
1,201-3,200			
3,201-10,000			
10,001-35,000	20	0	1
35,001-150,000	80	1	2
150,001-500,000	80	1	2

Single Sampling Plan for 1 Percent AQL Level S-3 Reduced

Lot size	Sample size	Accept	Reject
2-8	5	0	1
9-15			
16-25			
26-50			
51-90			
91-150			
151-280			
281-500			
501-1,200			
1,201-3,200			
3,201-10,000			
10,001-35,000	5	0	1
35,001-150,000	20	0*	2
150,001-500,000	20	0	2

*If the number of defects exceeds the accept number but is less than the reject number, accept the lot and return to normal inspection.

Single Sampling Plan for 2.5 Percent AQL Level S-3 Normal

Lot size	Sample size	Accept	Reject
2-8	5	0	1
9-15			
16-25			
26-50			
51-90			
91-150			
151-280			
281-500	5	0	1
501-1,200	20	1	2
1,201-3,200	20	1	2
3,201-10,000	20	1	2
10,001-35,000	20	1	2
35,001-150,000	32	2	3
150,001-500,000	32	2	3

Single Sampling Plan for 2.5 Percent AQL Level S-3 Tightened

Lot size	Sample size	Accept	Reject
2-8	8	0	1
9-15			
16-25			
26-50			
51-90			
91-150			
151-280			
281-500	8	0	1
501-1,200	32	1	2
1,201-3,200			
3,201-10,000			
10,001-35,000			
35,001-150,000			
150,001-500,000	32	1	2

Single Sampling Plan for 2.5 Percent AQL Level S-3 Reduced

Lot size	Sample size	Accept	Reject
2-8	2	0	1
9-15			
16-25			
26-50			
51-90			
91-150			
151-280			
281-500	2	0	1
501-1,200	8	0*	2
1,201-3,200	8	0	2
3,201-10,000	8	0	2
10,001-35,000	8	0	2
35,001-150,000	13	1	2
150,001-500,000	13	1	2

*If the number of defects exceeds the accept number but is less than the reject number, accept the lot and return to normal inspection.

Double Sampling Plan for 1 Percent AQL Level II Normal

Lot size	Sample size First	Second	Ac1	Ac2	Re1	Re2
2-8						
9-15						
16-25			Use single sampling plan			
26-50						
51-90						
91-150			Use single sampling plan			
151-280	32	32	0	2	1	2
281-500	32	32	0	2	1	2
501-1,200	50	50	0	3	3	4
1,201-3,200	80	80	1	4	4	5
3,201-10,000	125	125	2	5	6	7
10,001-35,000	200	200	3	7	8	9
35,001-150,000	315	315	5	9	12	13
150,001-500,000	500	500	7	11	18	19

Double Sampling Plan for 1 Percent AQL Level II Tightened

Lot size	Sample size First	Second	Ac1	Ac2	Re1	Re2
2-8						
9-15						
16-25			Use single sampling plan			
26-50						
51-90						
91-150			Use single sampling plan			
151-280	50	50	0	2	1	2
281-500	50	50	0	2	1	2
501-1,200	50	50	0	2	1	2
1,201-3,200	80	80	0	3	3	4
3,201-10,000	125	125	1	4	4	5
10,001-35,000	200	200	2	5	6	7
35,001-150,000	315	315	3	7	11	15
150,001-500,000	500	500	6	10	15	16

53

Double Sampling Plan for 1 Percent AQL Level II Reduced

Lot size	Sample size First	Second	Ac1	Ac2	Re1	Re2
2-8			Use single sampling plan			
9-15						
16-25						
26-50						
51-90						
91-150			Use single sampling plan			
151-280	13	13	0	2	0	2*
281-500	13	13	0	2	0	2
501-1,200	20	20	0	3	0	4
1,201-3,200	32	32	0	4	1	5
3,201-10,000	50	50	0	4	3	6
10,001-35,000	80	80	1	5	4	7
35,001-150,000	125	125	2	7	6	9
150,001-500,000	200	200	3	8	8	12

*If the number of defects exceeds the accept number but is less than the reject number, accept the lot and return to normal inspection.

Double Sampling Plan for 4 Percent AQL Level II Normal

Lot size	Sample size First	Second	Ac1	Ac2	Re1	Re2
2-8			Use single sampling plan			
9-15			Use single sampling plan			
16-25			Use single sampling plan			
26-50	8	8	0	2	1	2
51-90	8	8	0	2	1	2
91-150	13	13	0	3	3	4
151-280	20	20	1	4	4	5
281-500	32	32	2	5	6	7
501-1,200	50	50	3	7	8	9
1,201-3,200	80	80	5	9	12	13
3,201-10,000	125	125	7	11	18	19
10,001-35,000	200	200	11	16	26	27
35,001-150,000	200	200	11	16	26	27
150,001-500,000	200	200	11	16	26	27

Double Sampling Plan for 4 Percent AQL Level II Tightened

Lot size	Sample size First	Second	Ac1	Ac2	Re1	Re2
2-8			Use single sampling plan			
9-15			Use single sampling plan			
16-25			Use single sampling plan			
26-50	13	13	0	2	1	2
51-90	13	13	0	2	1	2
91-150	13	13	0	2	1	2
151-280	20	20	0	3	3	4
281-500	32	32	1	4	4	5
501-1,200	50	50	2	5	6	7
1,201-3,200	80	80	3	7	11	12
3,201-10,000	125	125	6	10	15	16
10,001-35,000	200	200	9	14	23	24
35,001-150,000	200	200	9	14	23	24
150,001-500,000	200	200	9	14	23	24

Double Sampling Plan for 4 Percent AQL Level II Reduced

Lot size	Sample size First	Second	Ac1	Ac2	Re1	Re2
2-8			Use single sampling plan			
9-15			Use single sampling plan			
16-25			Use single sampling plan			
26-50	3	3	0	2	0	2*
51-90	3	3	0	2	0	2
91-150	5	5	0	3	0	4
151-280	8	8	0	4	1	5
281-500	13	13	0	4	3	6
501-1,200	20	20	1	5	4	7
1,201-3,200	32	32	2	7	6	9
3,201-10,000	50	50	3	8	8	12
10,001-35,000	80	80	5	10	12	16
35,001-150,000	80	80	5	10	12	16
150,001-500,000	80	80	5	10	12	16

*If the number of defects exceeds the accept number but is less than the reject number, accept the lot and return to normal inspection.

Double Sampling Plan for 1 Percent AQL Level S-3 Normal

Lot size	Sample size		Ac1	Ac2	Re1	Re2
	First	Second				
2-8			Use single sampling plan			
9-15						
16-25						
26-50						
51-90						
91-150						
151-280						
281-500						
501-1,200						
1,201-3,200						
3,201-10,000						
10,001-35,000			Use single sampling plan			
35,001-150,000	32	32	0	2	1	2
150,001-500,000	32	32	0	2	1	2

Double Sampling Plan for 1 Percent AQL Level S-3 Tightened

Lot size	Sample size		Ac1	Ac2	Re1	Re2
	First	Second				
2-8			Use single sampling plan			
9-15						
16-25						
26-50						
51-90						
91-150						
151-280						
281-500						
501-1,200						
1,201-3,200						
3,201-10,000						
10,001-35,000			Use single sampling plan			
35,001-150,000	50	50	0	2	1	2
150,001-500,000	50	50	0	2	1	2

Double Sampling Plan for 1 Percent AQL Level S-3 Reduced

Lot size	Sample size		Ac1	Ac2	Re1	Re2
	First	Second				
2-8			Use single sampling plan			
9-15						
16-25						
26-50						
51-90						
91-150						
151-280						
281-500						
501-1,200						
1,201-3,200						
3,201-10,000						
10,001-35,000			Use single sampling plan			
35,001-150,000	13	13	0	2	0	2*
150,001-500,000	13	13	0	2	0	2

*If the number of defects exceeds the accept number but is less than the reject number, accept the lot and return to normal inspection.

Double Sampling Plan for 2.5 Percent AQL Level S-3 Normal

Lot size	Sample size		Ac1	Ac2	Re1	Re2
	First	Second				
2-8			Use single sampling plan			
9-15						
16-25						
26-50						
51-90						
91-150						
151-280						
281-500			Use single sampling plan			
501-1,200	13	13	0	2	1	2
1,201-3,200	13	13	0	2	1	2
3,201-10,000	13	13	0	2	1	2
10,001-35,000	13	13	0	2	1	2
35,001-150,000	20	20	0	3	3	4
150,001-500,000	20	20	0	3	3	4

Double Sampling Plan for 2.5 Percent AQL Level S-3 Tightened

Lot size	Sample size		Ac1	Ac2	Re1	Re2
	First	Second				
2-8			Use single sampling plan			
9-15						
16-25						
26-50						
51-90						
91-150						
151-280						
281-500			Use single sampling plan			
501-1,200	20	20	0	2	1	2
1,201-3,200						
3,201-10,000						
10,001-35,000						
35,001-150,000						
150,001-500,000	20	20	0	2	1	2

Double Sampling Plan for 2.5 Percent AQL Level S-3 Reduced

Lot size	Sample size		Ac1	Ac2	Re1	Re2
	First	Second				
2-8			Use single sampling plan			
9-15						
16-25						
26-50						
51-90						
91-150						
151-280						
281-500			Use single sampling plan			
501-1,200	5	5	0	2	0	2*
1,201-3,200	5	5	0	2	0	2
3,201-10,000	5	5	0	2	0	2
10,001-35,000	5	5	0	2	0	2
35,001-150,000	8	8	0	3	0	4
150,001-500,000	8	8	0	3	0	4

*If the number of defects exceeds the accept number but is less than the reject number, accept the lot and return to normal inspection.

BIBLIOGRAPHY

Juran, J.M., F.M. Gryna, and R.S. Bingham. *Quality Control Handbook,* 3rd ed. New York: McGraw-Hill Book Co., 1974.

Military Standard, MIL-STD-105D. "Sampling Procedures and Tables for Inspection by Attributes." Revision D, July 18, 1961.

Military Standard, MIL-STD-105E. "Sampling Procedures and Tables for Inspection by Attributes." Revision E, May 10, 1989.

Military Standard, MIL-STD-109. "QA Terms and Definitions." April 4, 1969.

Western Electric. *Statistical Process Control Handbook.* Easton, PA: Western Electric Co., 1956.

INDEX